Public Library District

JAN -- 2015

FARM ANIMALS
CHICKEN

Katie Dicker

A+
Smart Apple Media

Published by Smart Apple Media,
an imprint of Black Rabbit Books
P.O. Box 3263, Mankato, Minnesota, 56002
www.blackrabbitbooks.com

U.S. publication copyright © 2014 Smart Apple Media. International copyright reserved in all countries.
No part of this book may be reproduced in any form without written permission from the publisher.

Printed in the United States of America,
at Corporate Graphics in North Mankato, Minnesota.

Designed by Hel James
Edited by Mary-Jane Wilkins

Library of Congress Cataloging-in-Publication Data

Dicker, Katie.
  Chicken / Katie Dicker.
     p. cm. -- (Farm animals)
  Includes bibliographical references and index.
  ISBN 978-1-62588-018-5
  1. Chickens--Juvenile literature. I. Title.
  SF487.5.D525 2014
  636.5--dc23
                           2013000056

Photo acknowledgements
l = left, r = right, t = top, b = bottom
title page Valentina_S/Shutterstock; page 3 Bas Meelker/Shutterstock; 4 Jupiterimages, 5 Stockphoto/both Thinkstock; 6 iStockphoto/Thinkstock; 7 S.Cooper Digital/Shutterstock; 8 bogdanhoda/Shutterstock; 9 Hemera/Thinkstock; 10 JP Chretien/Shutterstock; 11 Digital Vision/Thinkstock; 12 Tumarkin Igor - ITPS/Shutterstock; 13 iStockphoto/Thinkstock; 14 Sue McDonald/Shutterstock; 15 iStockphoto/Thinkstock; 16 Thinkstock; 17 iStockphoto/Thinkstock; 18 iStockphoto/Thinkstock; 19 Comstock Images/Thinkstock, l Evgeny Semenov, b siamionau pavel/both Shutterstock; 20 l marilyn barbone/Shutterstock, r iStockphoto/Thinkstock; 21 t iStockphoto/Thinkstock, r and b Hemera/Thinkstock; 22 sanddebeautheil/Shutterstock; 23 Picsfive/Shutterstock
Cover Denis Nata/Shutterstock

DAD0507
052013
9 8 7 6 5 4 3 2 1

Contents

My World 4

Head to Toe 6

Time to Eat 8

Roosters and Hens 10

Laying Eggs 12

Hens and Chicks 14

Who Looks After Us? 16

Farm Produce 18

Chickens Around the World 20

Did You Know? 22

Useful Words 23

Index and Web Links 24

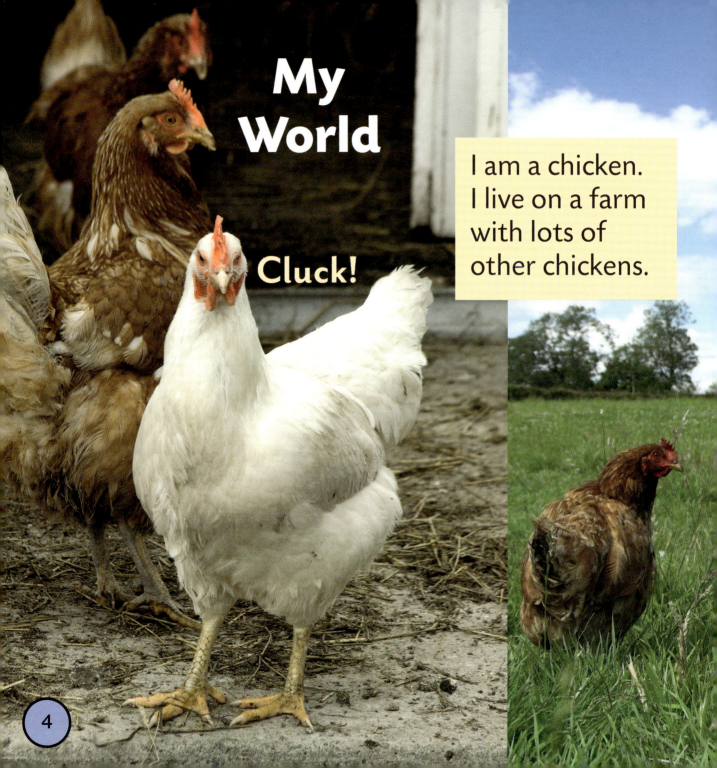

My World

Cluck!

I am a chicken. I live on a farm with lots of other chickens.

Here are some of my flock.
We like to live outside, but we go indoors at night to shelter and stay safe.

Chickens need to be kept safe from animals such as foxes and owls.

Head to Toe

On the top of my head is a comb.

Under my neck are two flaps of skin called wattles.

My feathers keep me warm and dry. I have feathery wings, but I cannot fly very far.

Chickens ruffle their feathers in the dust and dirt to keep their skin healthy.

Time to Eat

Peck

We like to eat seeds, leaves, grubs, and insects. The farmer also gives us grain.

We use our sharp claws to scratch the ground. Sometimes we find worms, snails, or slugs.

A chicken has small stones in its stomach to grind up its food.

Roosters and Hens

A male chicken is called a rooster. It has a large comb and colorful feathers to attract females.

Cock-a-doodle-doo!

Early in the morning, I make a loud crowing sound to warn other roosters to stay away.

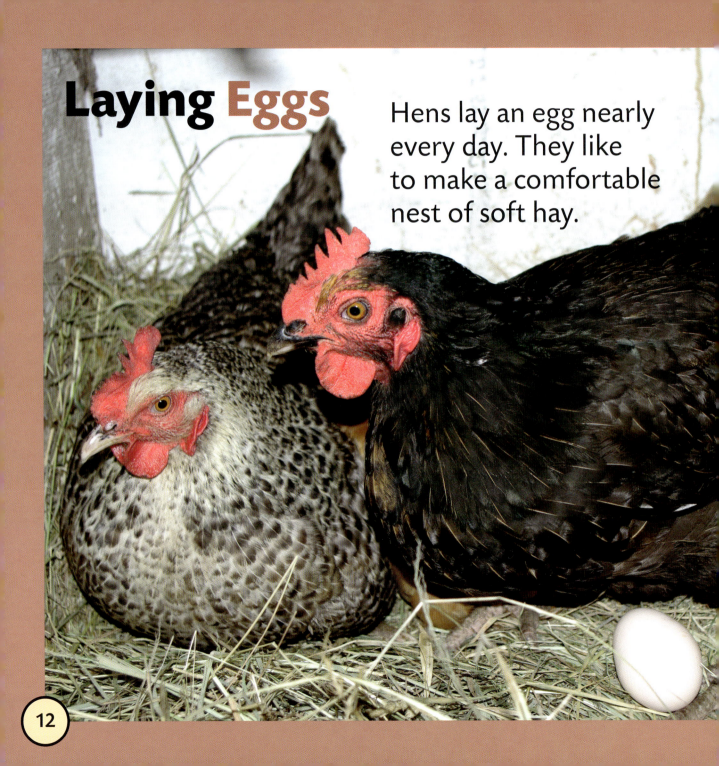

Laying Eggs

Hens lay an egg nearly every day. They like to make a comfortable nest of soft hay.

The eggs are usually pale brown or white, but some chicken eggs are light blue, green, or pink.

Hens and Chicks

Sometimes, a rooster is used to breed baby chicks. Each chick grows inside a hen's egg.

Chicks can run around and feed themselves soon after they hatch.

Cheep!

Chirp!

A hen can hear her chicks chirping. She gently clucks as they break out of the shell.

A chick has a tooth-like point on its beak to help it break out of the egg.

Who Looks After Us?

The farmer cleans our coop and gives us fresh hay to make our nests.

The farmer also checks we have clean water to drink, and collects our eggs every day.

Hens need to drink lots of water to keep laying eggs.

Farm Produce

Chickens are farmed for their meat or for their eggs.

Chickens farmed for their meat are called broiler chickens. When they reach a healthy weight, the farmer takes them to market.

Inside an egg there are two parts—the egg white and the yolk.

You can make lots of different food with eggs, such as omelets, meringues, cakes, and pancakes.

Chickens Around the World

Rhode Island Red, USA

Farmers in countries all over the world keep chickens, and many chickens are kept as pets. Here are some of the different breeds.

Bantam, Indonesia

Silkie, Asia

Silver Dorking, Italy

Buff Orpington, England

There are more than 150 different types of chickens around the world.

Did You Know?

Chickens usually live for about five to seven years.

Chickens have no teeth, but they use their sharp beaks to scoop up food and water.

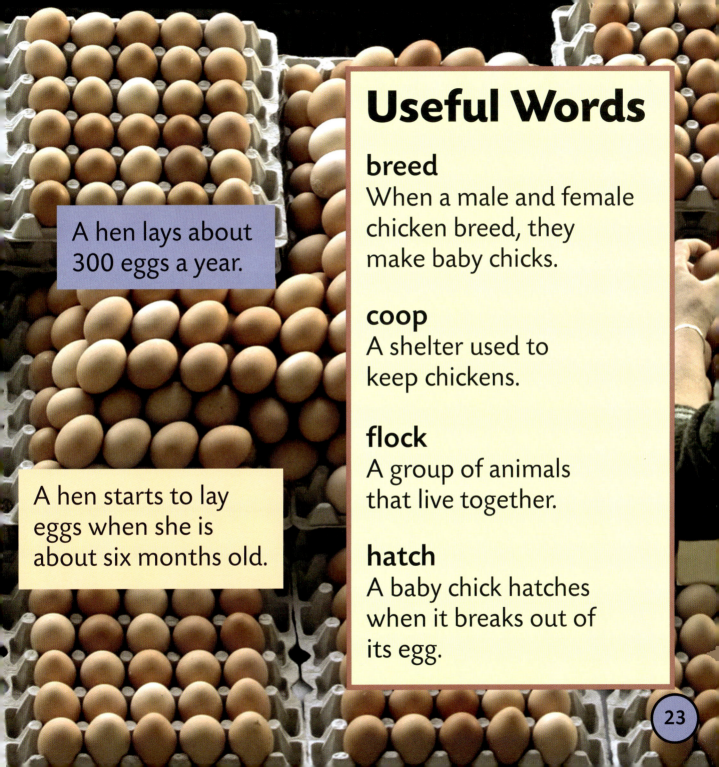

A hen lays about 300 eggs a year.

A hen starts to lay eggs when she is about six months old.

Useful Words

breed
When a male and female chicken breed, they make baby chicks.

coop
A shelter used to keep chickens.

flock
A group of animals that live together.

hatch
A baby chick hatches when it breaks out of its egg.

Index

beak 15, 22
breeds 20, 21

chicks 14, 15
claws 9
comb 6, 10
coop 16, 23
crowing 11

eating 8, 9, 14, 22

eggs 12, 13, 14, 15, 17, 18, 19, 23

feathers 7, 10
flock 5, 23

rooster 10, 11, 14

wattles 6
wings 7

Web Links

www.animalcorner.co.uk/farm/chickens/chicken_about.html
www.kidcyber.com.au/topics/farmpoultry.htm
www.ncagr.gov/cyber/kidswrld/general/barnyard/poultry.htm
www.kiddyhouse.com/Farm/Chicken
http://library.thinkquest.org/J0111462